A fascinating and amusing portrait of tl

Did you know, for instance, that the Ostrich holds many world records:

- It is the world's <u>largest bird</u>.
- It can run <u>faster</u> than any other two legged animal.
- This flightless bird lays the <u>largest eggs</u> of any living creature.

   - *Read inside for more amazing and incredible facts*
   - *Full of large pictures also suitable for colouring*

Hello, I'm Ozzy the Ostrich. See if you can find me in the rest of the book as I am hiding in 24 different places !

This book belongs to :

Acknowledgements: For my family and friends to whom I owe grateful thanks for all their encouragement, support and time in compiling this book. Special thanks go to my children, Francesca, Sebastian and Stephanie who kindly provided some of the illustrations, also to David for the use of his computer. Last, but not least, thanks to the ostriches for giving me such amusing and inspirational subject matter to work on.

Did you know it takes an *amazing* <u>two hours</u> to hard boil an ostrich egg ?

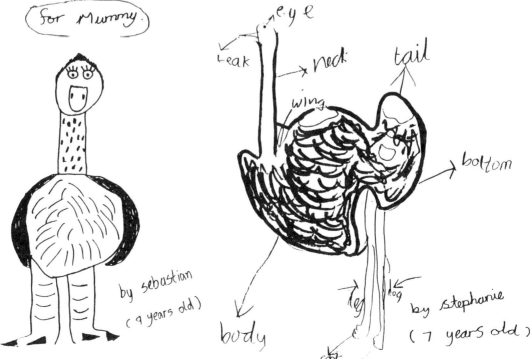

Contents

Feeding

Skeleton / Plumage

Activities

In the Wild

Eggs

At the End

1. **The word 'Ostrich' :** <u>What it means</u>

The ostrich is called **<u>camelus</u>** after the Arab name for the ostrich - **'camel bird'** because its profile at a distance is very similar to that of a camel.
It is also a member of the **<u>ratite</u>** family which means **'flightless'** bird.

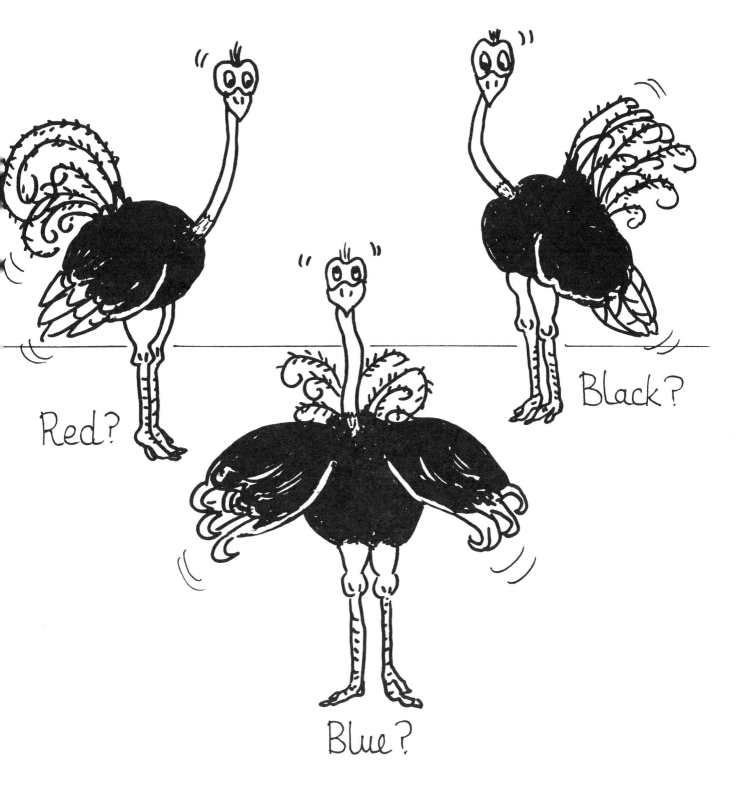

Red?

Black?

Blue?

2. **<u>Types of Ostrich</u> :** <u>Origin</u>

There are three basic types of ostrich - red-neck, blue-neck and the domesticated African Black. Ostriches originated from Africa and live mainly on grassland or semi-desert when living in their natural habitat. Domesticated ostriches have been farmed in Africa for over 100 years.

3. **In the Beginning :** Fossils

Seven million year old ostrich fossils have been found in rocks in southern Russia, India and China. It was thought that these ancestors could fly.

4. **Eggs :** <u>Incubation</u>

A normal ostrich egg takes about 42 days to incubate.
A baby ostrich chick has a speckled neck and, within a
month of hatching, it can run fast and feed itself.

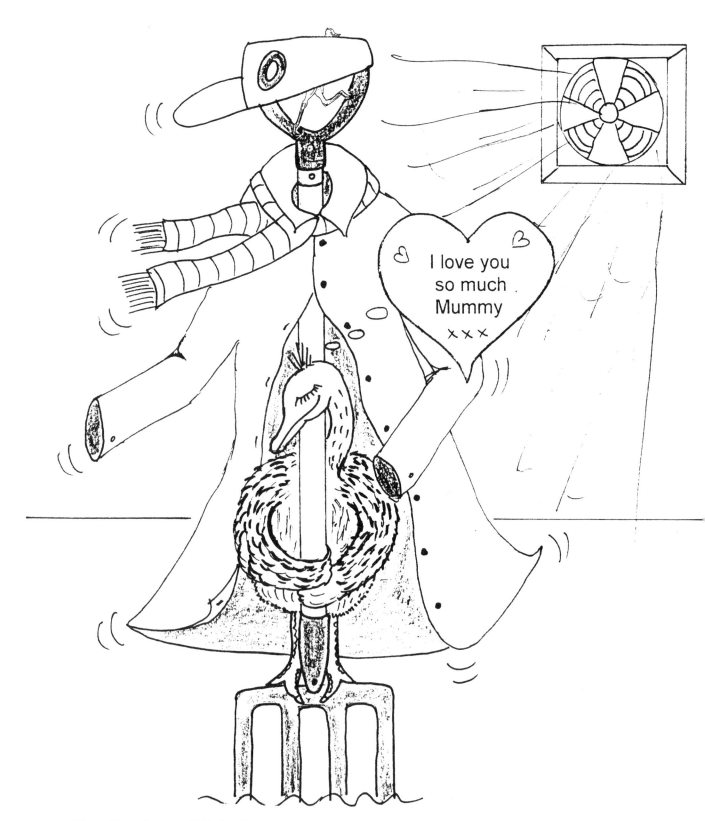

5. **Baby Chick :** <u>Mother Figure</u>

During the first two days after a baby chick has hatched out of its egg, the first object it sees (like a larger ostrich, person or animal which moves like a real mother) it will adopt as its mother. From then on, the adopted mother should always wear the same kind of clothes. Otherwise, a dummy wearing the same clothes should be placed in the chick pen, near ventilation to provide movement, and so fool the chick into thinking that 'mother' is still with them !

14 months old
= 8½ feet tall
Keep on eating your greens !!
6 months old
= 6 feet tall
4 months old
= 4 feet tall
2 months old
= 2 feet tall
1 month old
= 1 foot tall

6. **Growth :** Rate of Growth

A baby ostrich, or **chick**, grows an *amazing* 7.5 cm (3") in a week - that is 30 cm (12") in a month. It grows at this rate for the first seven months until it almost reaches adult height at 12 months of age, but is only 80 % of adult weight. Average heights are as follows :

1 month old	**=**	**30 cm**	**(12")**	**6 months old**	**=**	**180 cm**	**(72")**
2 months old	**=**	**60 cm**	**(24")**	**12 months old**	**=**	**240 cm**	**(96")**
4 months old	**=**	**120 cm**	**(48")**	**14 months old**	**=**	**255 cm**	**(102")**

7. __Height__ : __Tallest__ Bird in the World

The ostrich is the world's <u>tallest</u> bird. An adult male
ostrich will reach an *amazing* 2.75 metres (8½ feet) in
height when full grown. The average height of a ceiling
in a modern house is only 2.3 metres (7½ feet) high.
A fully grown bird stands nearly 1.5 metres (5 feet) high
at the top of its back.

8. **Size** : **Biggest** Bird in the World

The ostrich is the world's <u>largest</u> bird and weighs up to
an *amazing* 150 kg (23½ stone) compared to a man
weighing 70 kg (11 stone). Some ostriches even
weigh up to 180 kg which is just over 28 stone !

9. **Size :** <u>Smallest Bird in the World</u>

Compare the size of an ostrich's head to one of the smallest birds in the world - the tiny bee hummingbird which measures only 5 cm (2") long from beak to tail. This bird only weighs 1.7 gm (1/17 of an ounce) compared to an ostrich which weighs an *amazing* 150,000 gms (5,300 ounces).

10. **Speed** : <u>Fastest</u> <u>two legged animal in the world</u>

An ostrich walks at a speed of 4 km/h (2½ mph) and runs at 30 km/h (18½ mph) but can sprint at an *amazing* speed of 70 km/h (43 mph) when frightened.

11. This means it can run <u>faster</u> than any other two legged animal in the world. Compare this to a man who can run at only 36 km/h (22 mph) in a short sprint! It can also keep running at 65 km/h (40 mph) for up to 30 minutes.

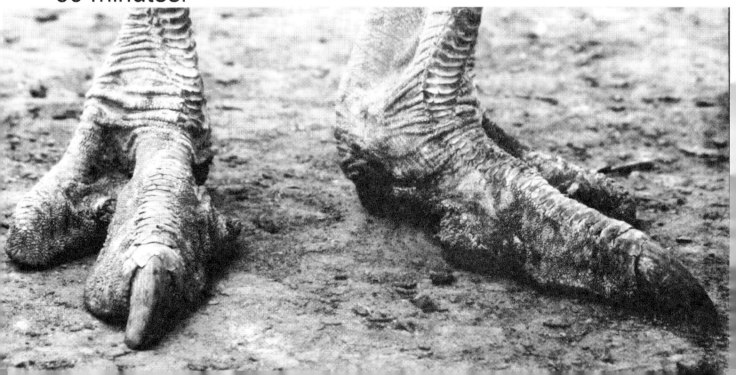

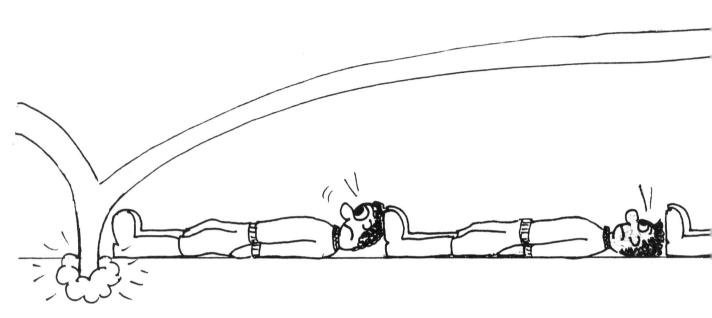

12. **Distance** : **Longest** stride

When an ostrich is running at full speed it can cover an *amazing* 8.4 metres (28 feet) in one single stride alone!

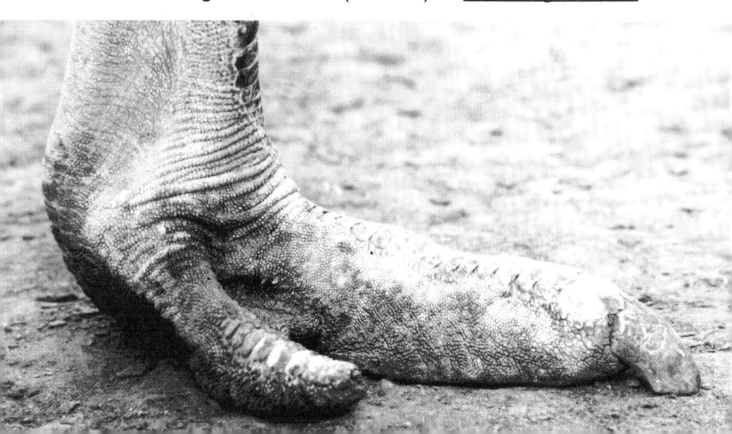

13. **Distance :** How many men ?

This means that an ostrich can leap over *almost* 5 men if they were all 1.8 metres (6 feet) tall !

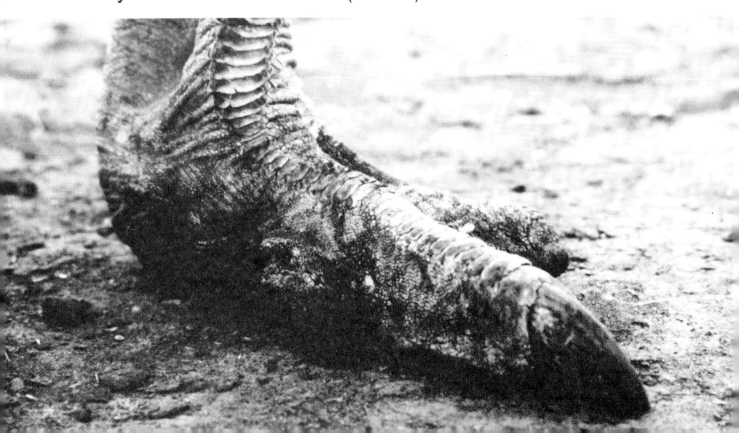

"**egg**strovert"   "**egg**scited"

14. <u>Legs</u> : <u>**Strongest** bird in the World</u>

The leg muscles of an ostrich are so strong that one kick can kill a lion or its strong toe nail can rip a man's stomach apart. Due to the structure and formation of an ostrich's leg it can only really kick forward.

15. **<u>Neck and Voice</u>** :   <u>How Long ? How Quiet ?</u>

The North African ostrich has a neck which is an *amazing*
1.1 metres (3½ feet / 42 ") long. The average man's neck is
only 10 cm (4 ") long.
An ostrich is very quiet as it does not have any vocal chords.
The voice box (**<u>syrinx</u>**) is so poorly developed that sounds
made vary from hoarse sighs to a little like the gentle 'yelp'
of a dog. However, when the male ostrich is in season and
is protecting its hens it can make a vibrating noise at the
back of its throat which sounds like the roar of a gentle lion.

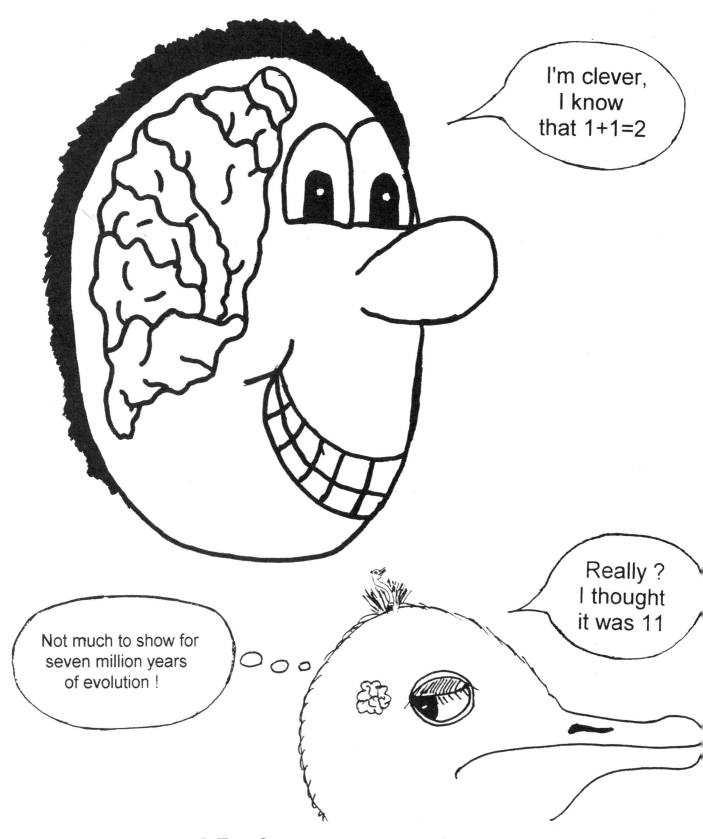

16. **Eyes and Brain :** Largest Eyes ? Smallest Brain ?

The **eyes** of an ostrich are the largest of any land animal in relation to the size of its head. They can focus on an object as far away as 5 km (3 miles). Believe it or not, the **brain** of an ostrich is only half the size of its eye. Look at ours !

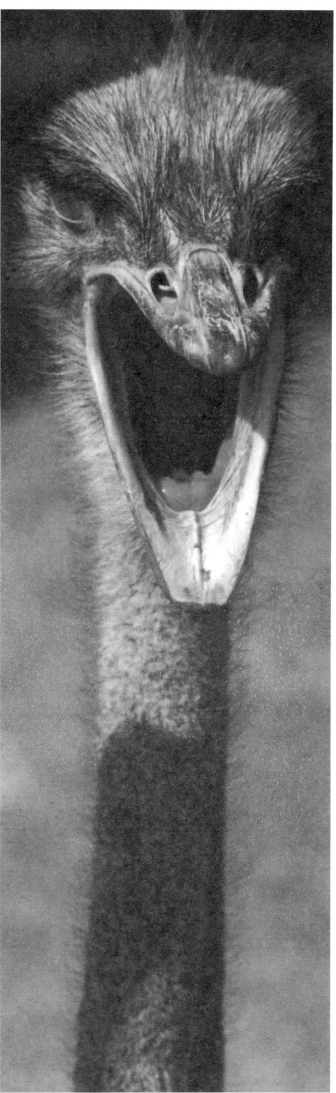

17. **Unusual Appetite :** <u>Bad Case of Indigestion</u>

One ostrich living in London Zoo, swallowed a spool of
film, 3 gloves, a comb, a bicycle valve, a pencil, some
rope, several coins, bits of a gold necklace, a collar
stud, a handkerchief and a clock.

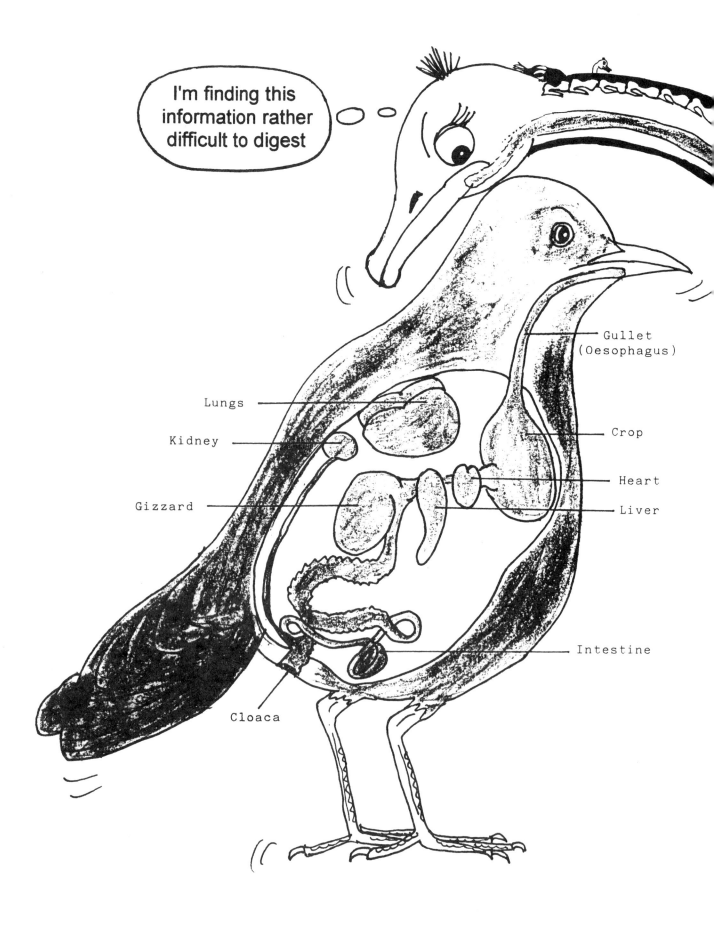

18. <u>**Inside a Bird**</u> :   <u>Digestive System</u>

Most birds have <u>two</u> stomachs but <u>not</u> the ostrich. For most birds the <u>first</u> stomach (the **<u>crop</u>**) stores the food and the <u>second</u> (**<u>gizzard</u>**) grinds it to a pulp with the help of grit and small stones.

(The large intestine is made up of the caecum (cecum), colon and rectum. The small intestine is made up of the duodenum, jejunum and ileum.)

Neck

Gullet
(Oesophagus)

Lungs

Kidney

Jejunum

Tail

Heart

Liver

Ventriculus
(Gizzard)

Proventriculus

Jejunum
Duodenum

Cecum

Cloaca

Rectum

Leg

19. **Inside an Ostrich** :   Digestive System

However, the ostrich does **not** have a crop. It has a **proventriculus** instead which is an organ where the food starts to be digested. This leads to the next part of the stomach called the **ventriculus** (gizzard) which contains small stones and grit which act as "the *teeth* of the ostrich" and which breaks down the food.

20. **Meal Times :** What an Ostrich Eats

Like many other birds, the ostrich needs to swallow grit and small stones with its food. In the wild, it also eats grass, seeds, leaves, flowers and occasionally locusts and grasshoppers. Domesticated ostriches, however, normally eat grass, alfalfa and specially prepared ostrich pellets. All ostriches need to drink water.

21. **Feeding** : <u>Eating Habits</u>

The ostrich lowers its long neck and pecks for its food. It
then stores this food in its gullet which is a pouch at the back
of its mouth, before finally lifting its head to swallow. The
food which accumulates in its gullet forms into a large ball
shape called a **bolus**. The passage of the bolus can be
clearly seen at it slides down inside the ostrich's stretchy
neck, down the food canal (oesophagus) until it eventually
arrives at its destination - the gut. A similar process is used
for drinking and a large ostrich can drink up to 5.7 litres
(10 pints) of water a day.

22. <u>**Problems with Digestion**</u> : <u>Impaction</u>

<u>**Impaction**</u> is a common reason for death in ostriches. This
is where the ostrich keeps eating food (like long grass)
before it has had time to be digested. The effect is like a
build-up in a blocked drain. The ostrich then starts to feel
uncomfortable (like us with indigestion) and stops eating and
drinking. One can try and get things moving again using
liquid paraffin but normally the ostrich dies from starvation
and dehydration three days later.

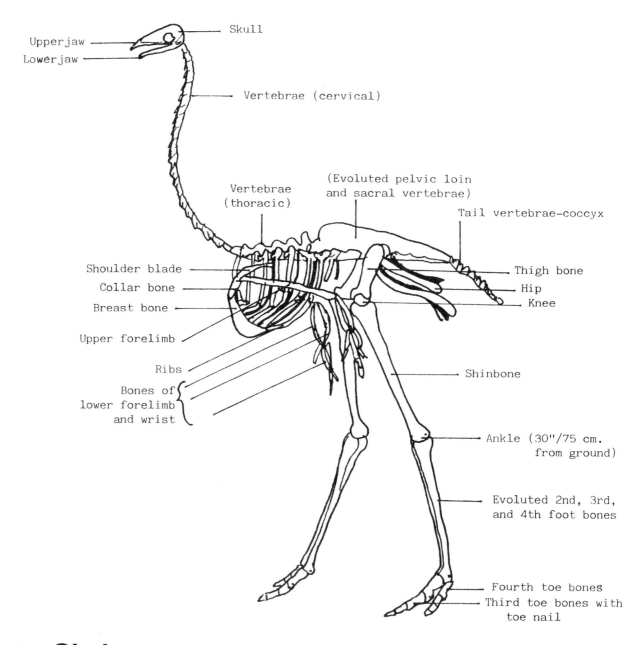

Upper jaw
Lower jaw
Skull
Vertebrae (cervical)
Vertebrae (thoracic)
(Evoluted pelvic loin and sacral vertebrae)
Tail vertebrae-coccyx
Shoulder blade
Collar bone
Breast bone
Upper forelimb
Ribs
Bones of lower forelimb and wrist
Thigh bone
Hip
Knee
Shinbone
Ankle (30"/75 cm. from ground)
Evoluted 2nd, 3rd, and 4th foot bones
Fourth toe bones
Third toe bones with toe nail

23. **Skeleton** : Main Parts

The wing skeleton of an ostrich is similar to that of flighted birds. However, as its wings are now too small and its feathers too soft for flying, it is unable to fly anymore. Instead, it is more adapted for walking and running. Two of the digits attached to its wings have claws but these are not readily seen due to feathering.

The ostrich is the **only** bird with **two** toes. The Emu and Cassowary have three but other birds have four. Some people believe that ostriches originally had five toes but only the third (largest) and fourth toes remain. Ostriches actually walk on their toes.

24. <u>**Plumage**</u> : <u>Colour of Feathers</u>

Chicks are born with buff colour spiky down with black tips. When the male ostrich is two years old it attains its full black and white plumage. The male's lower neck, body and entire wing coverts are black whereas the female is a dull grey brown colour. Most males have a collar of white feathers and white wing and tail feathers. Ostriches moult all year round and it takes 6 months for a wing feather to mature. Feathers grow up to three-quarters of a centimetre per day.

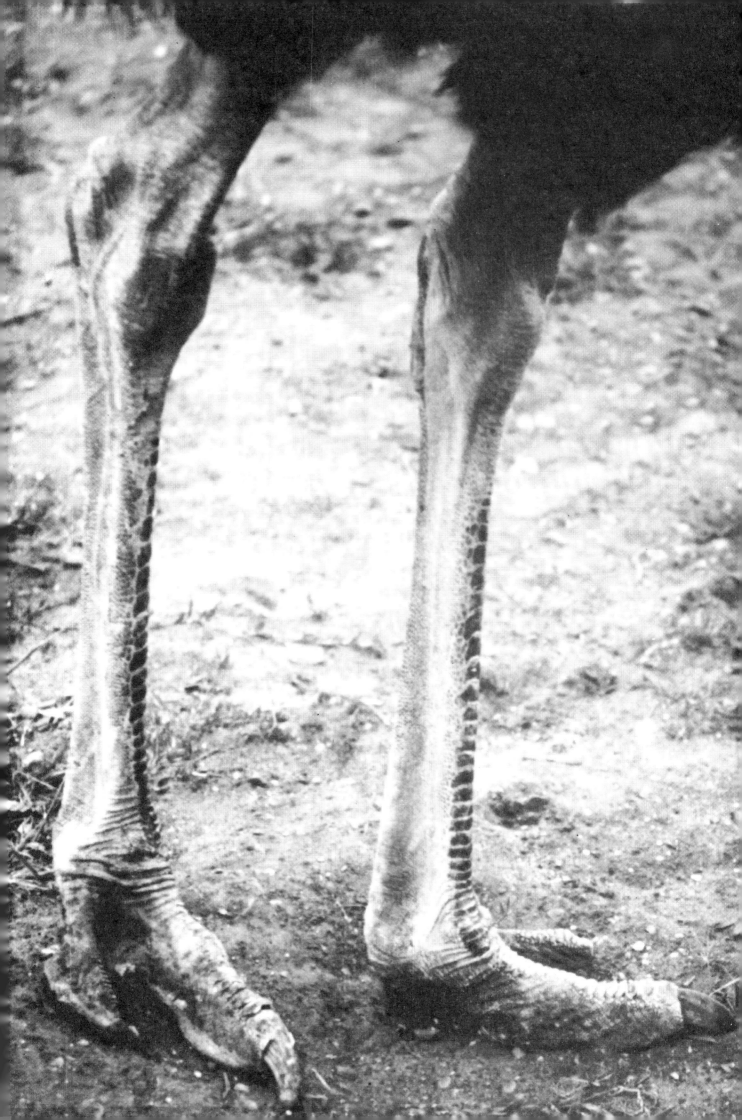

25. **Catching and Training Ostriches** : Unusual Duties

Ostriches have been trained to scare birds away from crops, to round up sheep as well as to be ridden in ostrich races. It is, however, forbidden by law, to ride ostriches in this country. Two items which are used to help catch the ostrich are a **crook** and a **hood**. The crook must be used carefully to avoid damaging the neck of the ostrich. The hood is probably the best means of quieting an ostrich and restraining it.

26. __Timespan :__ <u>For Breeding and Living</u>

A male ostrich reaches sexual maturity at about 3 years and a female at about 2 years. When a male ostrich is in season (ready to mate) his legs and neck become a bright pink and so does his beak (which looks rather like lipstick has been applied). They can breed up to an *amazing* 42 years of age and could produce more than 4,000 offspring in their lifetime! Ostriches have been known to live for up to 80 years of age so think twice before you buy one for a present!

DO NOT DISTURB

27. **Sleeping Habits** : Sleeping and Waking

The ostrich usually goes to sleep at sunset and wakes at sunrise. In captivity it is happy to sleep on a bed of straw. It usually sleeps in groups with its head and neck flat on the floor or slumped over the body of one of its friends, huddled together to keep warm. Baby chicks need 8 to 12 hours sleep each night.

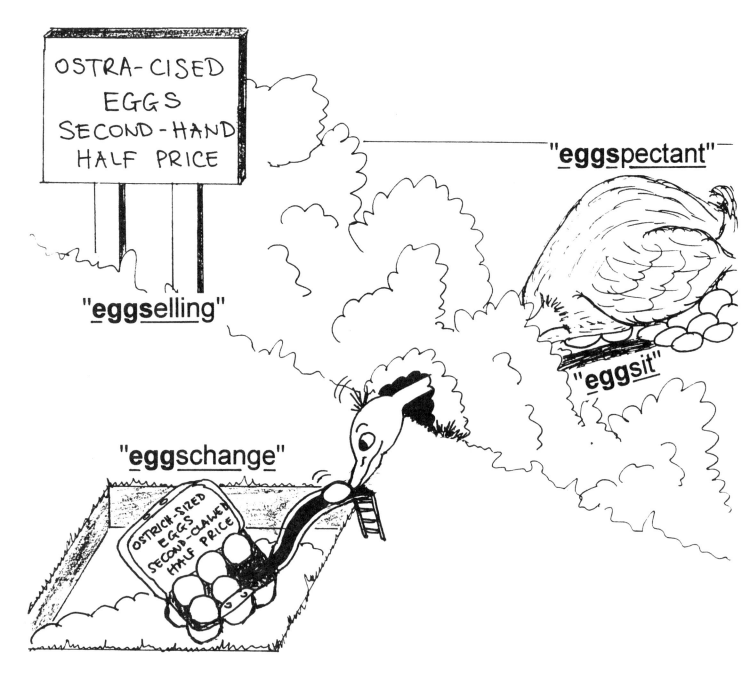

"eggspectant"

"eggselling"

"eggsit"

"eggschange"

OSTRA-CISED EGGS SECOND-HAND HALF PRICE

OSTRICH-SIZED EGGS SECOND-CLAWED HALF PRICE

28. **In the Wild :** Nests

In the wild, the male ostrich or **rooster** makes several shallow nest **scrapes**. He pairs up with a **major hen** who then lays up to eleven eggs in her selected scrape. In the same nest, two to five **minor hens** also lay their eggs . The major hen incubates the eggs during the day, the male at night. The ostrich can incubate 20 eggs. However, if there are too many eggs (up to 40 !) in the same nest, the major hen will roll some of the minor hens eggs to the edge to make sure they won't hatch.

Editors Note: There were so many *eggs*amples, I couldn't find space for them all. Where would you put these:- *eggs*iled, *eggs*panding, *eggs*pendable, *eggs*clusive, *eggs*ecute, *eggs*terminate, *eggs*porting, *eggs*odus, *eggs*tracting, *eggs*traditing ? Can you think of any more ? See 30 to 37 for more *eggs*amples.

29. **In the Wild :** Guarding Chicks / Head in Sand

Only 15 % of ostrich chicks hatched in the wild will survive to
become adults as the others normally fall prey to hyenas
and jackals. The male ostrich spreads out its wings and
kicks with its strong legs to defend his chicks against
predators while the female ostrich stays back to guard them.
It is a common misconception that the ostrich buries its head
in the sand. This legend probably came about because,
from a distance, the tiny head of a grazing ostrich may not
be clearly noticed.

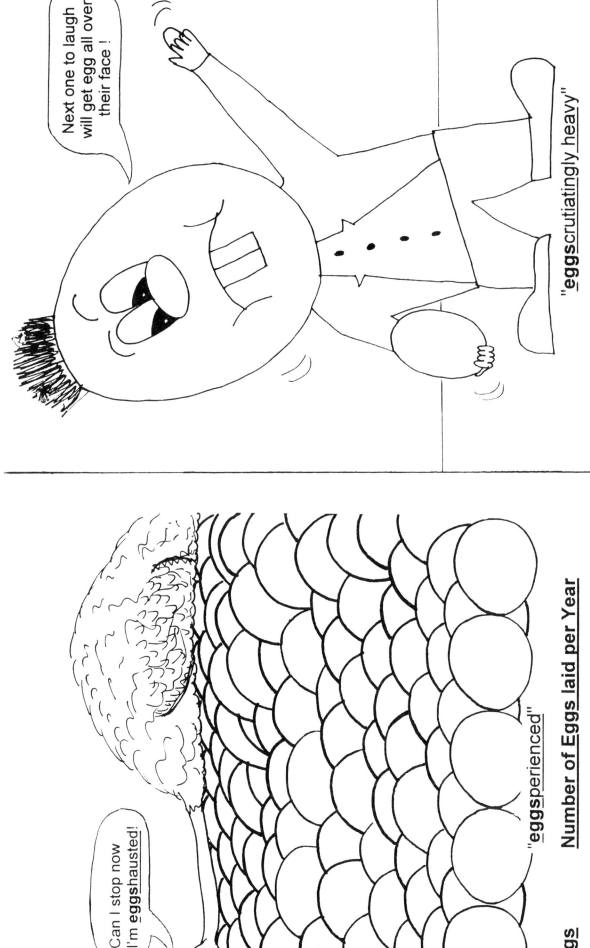

"eggsperienced"

30. Eggs

Number of Eggs laid per Year

An adult female ostrich can lay an *amazing* 100 eggs or more in one year. Remarkably, one African Black ostrich in the UK laid over 135 eggs between February and September of 1993 of which 80 % were fertile. Can you count how many eggs this hen has laid ?

"eggscrutiatingly heavy"

31. Eggs

Largest Eggs in the World

The ostrich lays the largest eggs of any living creature. Their eggs weigh up to 1.9 kg (4 lbs / 64 oz) compared to an average hen's egg which weighs only 55 gms (2 oz). However, an ostrich egg is the *smallest* egg in relation to the eventual size of the fully grown bird.

"**eggs**tremely **s**mall"

"**eggs**aggerated"

33. Eggs

Length

An ostrich egg is about 20 cm (8") long compared to a hen's egg which is only 5 cm (2") long. The ostrich egg is an *amazing* 30 times heavier than a hen's egg. The eggs have a volume of about 1.7 litres (3 pints).

At least you had a cracking time

"**eggs**ploded"

"**egg**ceptional"

32. Eggs

Strongest eggs in the world

The ostrich lays the strongest egg in the world. The egg is so strong (its shell is 0.25 cm (1/10") thick) that it won't break even if a man stands on it. Some African tribes use ostrich eggshells as water containers.

43

"**egg**shilarated"

"**egg**asperated"

How many omelettes ?

For Cooking

35. **Eggs**

Think of the time you would save if you only had to break open one egg instead of 25 in order to make about 12 omelettes.

34. **Eggs**

One ostrich egg contains the equivalent of 25 hen's eggs in volume.

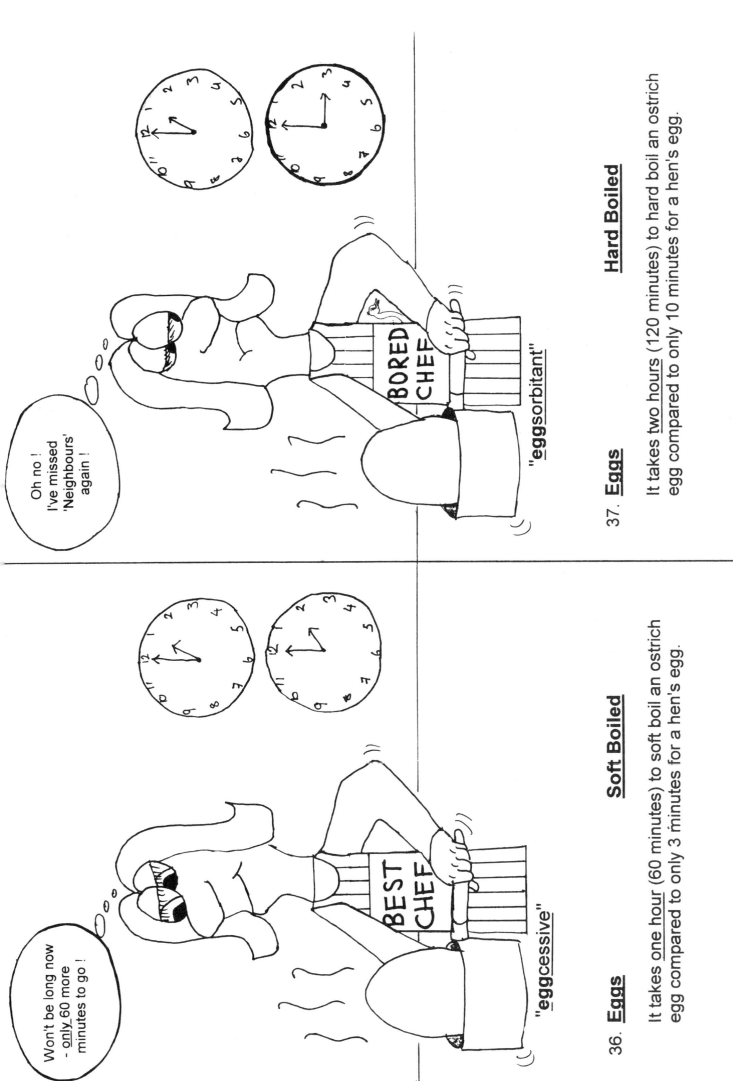

"eggsorbitant"

Hard Boiled

37. **Eggs**

It takes <u>two hours</u> (120 minutes) to hard boil an ostrich egg compared to only 10 minutes for a hen's egg.

"eggcessive"

Soft Boiled

36. **Eggs**

It takes <u>one hour</u> (60 minutes) to soft boil an ostrich egg compared to only 3 minutes for a hen's egg.

38. Commercial Uses of the Ostrich

All parts of an ostrich can be used effectively, for example:

1. Its meat (also known as 'volaise'). Many people believe ostrich meat is a white meat rather like that of a turkey. It is, however, red in colour and is very similar to beef in both taste and texture. It is a very healthy meat for us to eat as it is low in calories, fat and cholesterol but full of protein. Many of the top athletes in South Africa eat ostrich meat for this reason. Ostrich meat is currently sold in a variety of ways: pastrami, frankfurters, ham, pate, bacon, fillet steaks, sun dried (called 'jerky' or 'biltong') and even ostrich burgers. There are 36 kilograms (80 lbs) of prime meat on the carcass of a 12 month old ostrich.

2. Its skin (hide) known as the 'Rolls Royce' of leathers because of its unusual quill mark pattern, is made into fashionable clothes such as jackets and trousers. The Princess of Wales recently bought a pair of ostrich skin trousers which cost an *amazing* £4,000. 1½ square metres (17 square feet) of valuable leather can come from just one 12 month old ostrich.

3. The skin from its leg is used for fashion accessories such as boots, shoes, handbags, purses, etc.

4. Its feathers are used as hat plumes or as oil free anti-static feather dusters especially for cleaning fine machinery such as computers.

5. Its infertile eggs are used for lamp shades, decorated ornaments or even water containers.

6. Other uses are oils for medicinal purposes.

39. Cooking with Ostrich Meat - Chef's Tips

(1) It is necessary to take the meat out of a fridge and to let it stand before cooking, for example, 15 minutes for small pieces (steaks, fillets) and one hour for larger cuts of meat.

(2) Never prick the meat.

(3) Remove any fat from the meat before cooking as it has a stronger smell and taste than the meat itself and can give the meat a stronger taste.

(4) As ostrich meat is a red meat it should be browned in a frying pan on a high heat for a few minutes to seal in the juices. Either use butter or olive oil and butter for frying.

(5) As ostrich meat does not contain much fat, care should be taken not to overcook it as the meat will become dry very quickly.

(6)   Ostrich meat steaks are usually cooked very quickly, and served rare. Melt the butter (or olive oil and butter) in a frying pan.   When this is hot, fry the ostrich steaks quickly for about 2 - 3 minutes on each side. (This is for a steak which is about 3 cm thick).

(7)   For larger cuts of meat, first cook it quickly as above, so as to seal in the juices, and then allow 15 minutes per 500 g of meat and cook in the oven at 180°C / 350°F or Gas Mark 4. Make sure you baste it several times during cooking.

(8)   After cooking has finished and before serving the meat, allow the meat to rest in a warm place - 10 minutes for steaks or fillets and 30 minutes for larger pieces.

40. Ostrich Steak with Brandy Cream Sauce

Ingredients: (for 4 people)

4 ostrich steaks (about 3 cm thick / 200 g (7 oz) each)
knob of butter
2 tablespoons of *Bristol Blend 5 peppers (crushed)

(*you can buy these already mixed in a jar at most Supermarkets. The mix is made up of black, green, white and pink peppercorns with pimento added to give a special piquancy.   If not available, use whole black peppercorns).

The Sauce:

1 tablespoon of French mustard   2 tablespoons of Brandy
250 ml (8 fl. oz.) single cream   salt
knob of butter

To Prepare:

Sprinkle salt on steaks and coat the steaks well with crushed peppers. Melt the butter in a frying pan, and when hot fry the steaks quickly for about 2 - 3 minutes on each side. Place steaks on serving dish and keep warm. Add Brandy to frying pan and stir to mix in with remaining juices.

The Sauce:

In this same frying pan, add the mustard and cream. Mix well and reduce as necessary to make a pouring sauce. Add the butter to melt and mix together.

To Serve: Serve with saute potatoes and salad or vegetable of your choice.

Variations: Masala/Madeira Sauce

EITHER   Fry steaks as above and place on serving dish to keep warm. Add three tablespoons of Masala/Madeira and mix with juices from pan. When hot pour over steaks and eat immediately.

OR   Fry ostrich steaks as above and place on a serving dish to keep warm.   Mix two tablespoons of Masala/Madeira and 250 ml (8 fl. oz.) stock/bouillon with juices in the frying pan. Bring to the boil and add 60 gms. (2 oz.) *beurre manie (*in a separate bowl mix 30 gms. (1 oz.) butter with 30 gms. (1 oz.) plain flour together to form a ball. Stir in the beurre manie to thicken sauce and pour onto meat and serve straight away.

41. Ostrich Meat - Its Advantages

1.   Ostrich meat is lower in cholesterol and fat than chicken, turkey, beef, pork or lamb.

2.   Not only is it lower in cholesterol and fat, it is also lower in calories but rich in protein.

3.   Weight for weight, ostrich meat has just a little over half the fat of chicken and about a ninth of that of beef steak. It has over 70% fewer calories than best beef and two-thirds of the cholesterol.

4.   As consumption of beef continues to decline, hit by the scare over BSE, people are seeing ostrich meat as a welcome healthy alternative.

5.   Not only is it a healthy alternative to beef but most recipes suitable for beef will work using ostrich meat, and so will standard accompaniments, like orange sauce (usually served with duck) or Masala sauce (usually served with venison).

6.   Ostrich meat is a red meat and very similar in looks and taste to beef but it tastes like the finest fillet (or possibly a little better - more finely textured and delicious).

7.   It can also be stir-fried and as it is very low in cholesterol and has virtually no fat, there is little shrinkage during cooking.

8.   Ostrich meat can be marinated in wine, used as a main ingredient in a stew, baked 'en croute' and fried using the 'cordon bleu' method - it is a very versatile meat.

42. Comparison of Meats

Type of Meat (based on 85 gm per portion)	Calories	Fat (gm)	Cholesterol (mg)
Ostrich	96.9	Trace	49.3
Pork	275	7.01	84
Beef (steak)	240	6.4	77
Lamb	205	5.6	78
Chicken	140	0.9	73
Turkey	135	0.9	59

Printed by Fieldfare Publications, Cambridge